AMAZING EVOLUTION

VOLUME 4: THE WILY HEDGE SPARROW, STRANGE-SHAPED BEAKS, FISHY TALES AND MORE

A Book in the Nature's Wonders series

Kerry Atkinson

About this book

Key words: Evolution, survival of the fittest, Charles Darwin, natural history, nature

Acknowledgements

I would like to acknowledge the help that I received from Jim Severino in designing the covers for this book and for uploading it for me to Kindle Direct Publishing.

Other books by Kerry Atkinson

Medical books

1. Graft-versus-Host Disease. Editors, S Burakoff, H J Deeg, J Ferrara, K Atkinson. Marcel Dekker Inc, New York, 1990.
2. Clinical Bone Marrow Transplantation. First edition. Editor, K Atkinson. Cambridge University Press, Cambridge UK, 1994.
3. The BMT Data Book: A Manual for Bone Marrow and Blood Stem Cell Transplantation. K. Atkinson, Cambridge University Press, Cambridge UK, 1998.
4. Clinical Bone Marrow and Blood Stem Cell Transplantation. Second Edition. Editor, K. Atkinson. Cambridge University Press, Cambridge UK, 2000.

5. Clinical Bone Marrow and Blood Stem Cell Transplantation. Third Edition. Eds, K. Atkinson, R Champlin, J Ritz, M Brenner, P Ljungman and W Fibbe. Cambridge University Press, Cambridge UK, 2004.
6. The BMT Data Book. Second edition. R Munker, HM Lazarus and K. Atkinson, Cambridge University Press, Cambridge UK, 2009.
7. The BMT Data Book: Including Cellular Therapy. Third edition. R Munker, C Hildebrandt, HM Lazarus and K. Atkinson, Cambridge University Press, Cambridge UK. 2013.
8. Hematopoietic Cell Transplants: Concepts, Controversies and Future Directions. Eds HM Lazarus, RP Gale, A Keating, A Bacigalupo, R Munker, K Atkinson, A Syed. Cambridge University Press, Cambridge, UK. 2017.
9. The Biology and Therapeutic Application of Mesenchymal Cells. Ed. K Atkinson. Wiley/Blackwell. 2017
10. Revolutionizing Tropical Medicine. Eds. K Atkinson and D Mabey. Wiley/Blackwell Publishers Inc. 2019.
11. Asthma: Targeted Biological Treatments 2020. Kindle Direct Publishing, 2020. e-book and paperback
12. Atopic Dermatitis: Targeted Biological Treatments 2020. Kindle Direct Publishing, 2020. e-book and paperback
13. Psoriasis: Targeted Biological Treatments 2020. Kindle Direct Publishing, 2020. e-book and paperback
14. Crohn's disease and ulcerative colitis: Targeted Biological Treatments 2020. Kindle Direct Publishing, 2020. e-book and paperback
15. Rheumatoid arthritis, Juvenile Idiopathic Arthritis, Ankylosing spondylitis, Psoriatic Arthritis and Gout: Targeted Biological Treatments 2020. Kindle Direct Publishing, 2020. e-book and paperback
16. Small Cell Lung Cancer: Targeted Biological Treatments 2020. Kindle Direct Publishing, 2020. e-book and paperback
17. Chronic Myeloid Leukemia: Targeted Biological Treatments 2020. Kindle Direct Publishing, 2020. e-book and paperback
18. Melanoma: Targeted Biological Treatments 2020, Kindle Direct Publishing 2020. e-book and paperback
19. Dementia - A Caregiver's Guide. Kindle Direct Publishing 2020. e-book

20. Breast Cancer: Targeted Biological Treatments 2021. Kindle Direct Publishing 2021. e-book and paperback
21. Non-melanoma Skin Cancers: Targeted Biological and Specifically Designed Therapeutic agents. 2022. Kindle Direct Publishing 2021. e-book and paperback
22. Hodgkin Lymphoma: Targeted Biological Treatments. 2022. Kindle Direct Publishing 2022. e-book and paperback
23. Chimeric Antigen Receptor (CAR) T Cells: The second Revolution in Cellular Therapy. 2022. Kindle Direct Publishing 2022. e-book and paperback
24. Acute lymphoblastic leukemia: targeted immune therapeutic agents. Kindle Direct Publishing 2023. e-book and paperback
25. Artificial Intelligence and Medicine. A book in the Modern Medicine series. Kindle Direct Publishing 2024. e-book and paperback
26. Artificial Intelligence in Medicine.Volume 2: Clinical Applications before the Availability of CHAT-GTP4. a book in the Modern Medicine series. Kindle Direct Publishing 2024. e-book and paperback
27. Check Point Inhibitors: a New Class of Drugs for Cancer Treatment. Kindle Direct Publishing 2024. e-book and paperback
28. Cancer treatment updates January 2024 – October 2024. Kindle Direct Publishing 2024. e-book and paperback

Natural history/conservation books

29. Bird facts for everyone. K Atkinson. Outskirts Press Inc., Denver, USA. 2014
30. Birds, Birds, Birds. Kindle Direct Publishing 2020. e-book and paperback
31. A Field Guide to the Wild Habitats of Sydney Australia. Kindle Direct Publishing 2021. e-book and paperback
32. Conserving New Zealand's Endangered Species. Kindle Direct Publishing 2021. e-book and paperback
33. Conserving Australia's Endangered Species. Kindle Direct Publishing 2021. e-book and paperback
34. Nature's Wonders: Britain and Ireland. Kindle Direct Publishing 2021. e-book and paperback.
35. Nature's Wonders: Spain and Bulgaria. Kindle Direct Publishing 2022. e-book and paperback.

36. Nature's Wonders: France's Camargue, the Swiss National Park and Portugal. Kindle Direct Publishing 2022. e-book and paperback.
37. Nature's Wonders: Cuba. Kindle Direct Publishing 2022. e-book and paperback.
38. Nature's Wonders: Japan in Winter Kindle Direct Publishing 2022. e-book and paperback.
39. Nature's Wonders: India. Kindle Direct Publishing 2022. e-book and paperback.
40. Nature's Wonders: Tanzania. Kindle Direct Publishing 2022. e-book and paperback.
41. The Silk Road: Uzbekistan and Kazakhstan – A Book in the Nature's Wonders Series. 2022. Kindle Direct Publishing 2022. e-book and paperback.
42. Madagascar – Land of Lemurs and Chameleons: a book in the Nature's Wonders series 2023. Kindle Direct Publishing 2022. e-book and paperback.
43. South Africa - Kalahari to the Cape - a book in the Nature's Wonders series. Kindle Direct Publishing 2022. e-book and paperback
44. Nature's Wonders – Finland - land of owls and lakes. Kindle Direct Publishing 2023. e-book and paperback
45. Nature's Wonders – Ethiopia. Kindle Direct Publishing 2023. e-book and paperback
46. Nature's Wonders – Kakadu National Park and Central Australia. Kindle Direct Publishing 2023. e-book and paperback
47. Nature's Wonders – Lord Howe Island, Australia. Kindle Direct Publishing 2023. e-book and paperback
48. Norway above the Arctic Circle in March. A book in the Nature's Wonders series. Kindle Direct Publishing 2024. e-book and paperback
49. The Inside Passage and the Pribilof Islands. A book in the Nature's Wonders series. Kindle Direct Publishing 2024. e-book and paperback
50. Orca! A book in the Nature's Wonders series. Kindle Direct Publishing 2024. e-book
51. Watching Humpback Whales from your Sofa, while sipping a Glass of Chardonnay (Australian of course) or…The Lazy Naturalist's Guide to Local Sydney Wildlife. Kindle Direct Publishing 2024. e-book and paperback
52. Magic Moments of an Amateur Naturalist. A book in the Nature's Wonder's series. Kindle Direct Publishing 2024. e-book and paperback

53. Amazing evolution Volume 1. Kindle Direct Publishing 2024. e-book and paperback
54. Nature's Wonders – the USA. Kindle Direct Publishing 2024. e-book and paperback
55. Amazing evolution Volume 2. Kindle Direct Publishing 2024. e-book and paperback
56. South Island, New Zealand. Kindle Direct Publishing 2024. e-book and paperback

Other books

57. Paintings by Kerry Atkinson 2019 – 2021. Kindle Direct Publishing 2021. e-book and paperback.
58. Ferrari Road Cars. Kindle Direct Publishing 2022. e-book and paperback.
59. Paintings by Kerry Atkinson. 2022. Kindle Direct Publishing 2022. e-book and paperback.
60. Samuel Beckett and Bob Dylan – living in a Shakespearean era. Kindle Direct Publishing 2022. e-book and paperback.
61. Paintings and sculptures by Kerry Atkinson 2023. Kindle Direct Publishing 2023. e-book and paperback.
62. Ten influential European paintings. Kindle Direct Publishing 2023. e-book and paperback
63. The Pillars of Rock 'N Roll. Kindle Direct Publishing 2024. e-book and paperback
64. Early 20th Century Western paintings - a smorgasbord of color and radical shapes. Kindle Direct Publishing 2024. e-book
65. Insects and other wildlife included in Dutch flower paintings from the Dutch golden age. Kindle Direct Publishing 2024. e-book and paperback
66. Paintings and sculptures by Kerry Atkinson 2024
67. Deserts of the World and their wildlife. Kindle Direct Publishing 2025. e-book and paperback
68. Mountain ranges of Central Asia and their wildlife. Kindle Direct Publishing 2025. e-book and paperback
69. Brisbane Birds. A book in the Nature's Wonders series. Kindle Direct Publishing 2025. e-book and paperback

70. Lake Baikal and its wildlife. A book in the Nature's Wonders series. Kindle Direct Publishing 2025. e-book and paperback
71. Little known seas of the World and their wildlife. Kindle Direct Publishing 2025. e-book and paperback

Other books

72. Paintings by Kerry Atkinson 2019 – 2021. Kindle Direct Publishing 2021. e-book and paperback.
73. Ferrari Road Cars. Kindle Direct Publishing 2022. e-book and paperback.
74. Paintings by Kerry Atkinson. 2022. Kindle Direct Publishing 2022. e-book and paperback.
75. Samuel Beckett and Bob Dylan – living in a Shakespearean era. Kindle Direct Publishing 2022. e-book and paperback.
76. and sculptures by Kerry Atkinson 2023. Kindle Direct Publishing 2023. e-book and paperback.
77. Ten influential European paintings. Kindle Direct Publishing 2023. e-book and paperback
78. The Pillars of Rock 'N Roll. Kindle Direct Publishing 2024. e-book and paperback
79. Early 20th Century Western paintings - a smorgasbord of color and radical shapes. Kindle Direct Publishing 2024. e-book
80. Insects and other wildlife included in Dutch flower paintings from the Dutch golden age. Kindle Direct Publishing 2024. e-book and paperback
81. Paintings and sculptures by Kerry Atkinson 2024
82. Cadillacs with Rocket-like Tailfins. Kindle Direct Publishing 2025. e-book and paperback
83. Boy Racer. Kindle Direct Publishing 2025. e-bo1. ok and paperback

Quotation from The Holy Modal Rounders

"If you want to be a bird

It won't take much

To get you up there

When you come down

Land on your feet"

From the song "If you want to be a bird" by the Holy Modal Rounders. Originally released on a 1968 Holy Modal Rounders album as "The Bird Song," this track was later included on the "Easy Rider" film soundtrack album, released by ABC-Dunhill Records in August 1969.

TABLE OF CONTENTS

Author's preface to Amazing Evolution Volume 1

Who is the World's greatest genius? Some might say Michelangelo for his painting of the ceiling of the Sistine Chapel and his sculpture of David. Some might say Mozart for his ethereal music. Some might say Albert Einstein for his theory of relativity. I would add another candidate – Charles Darwin (1809 – 1882) for his explanation of the evolution of living creatures. When visiting the Galapagos Islands on HMS Beagle, Darwin noted that finches on one island had differently shaped beaks to finches on another nearby island. This enabled them to access the nectar of flowers that their rivals on the other island could not reach. Eureka moment! This adaptation was useful in terms of their ability to prosper and was incorporated into their DNA and thus the gene passed on to their offspring: survival of the fittest or natural selection.

The planet provided the necessary molecules for life – carbon, hydrogen, oxygen and methane. From the humblest single cell organism, more complex organisms were derived – bacteria, plankton, plants, insects, fish, reptiles, amphibians and mammals. This book describes some of the most striking examples of evolution.

Author's preface to Amazing Evolution Volume 1

Chales Darwin as a boy

This image is by Ellen Sharples - File originally uploaded to w:Wikipedia by Duncharris at 20:49, 5 May 2004., Public Domain, https://commons.wikimedia.org/w/index.php?curid=36177; n.wikipedia.org/wiki/Charles_Darwin#/media/File:Charles_Darwin_1816.jpg. This work is in the public domain in its country of origin and other countries and areas where the copyright term is the author's life plus 100 years or fewer. PD 1925. No changes were made to this image.

Charles Darwin as a young man

This image is by George Richmond - From Origins, Richard Leakey and Roger Lewin, Public Domain, https://commons.wikimedia.org/w/index.php?curid=22092879; By George Richmond - From Origins, Richard Leakey and Roger Lewin, Public Domain, https://commons.wikimedia.org/w/index.php?curid=22092879. This work is in the public domain in its country of origin and other countries and areas where the copyright term is the author's life plus 100 years or fewer. PD 1925. No changes were made to this image.

Darwin's finches

1. Geospiza magnirostris.
2. Geospiza fortis.
3. Geospiza parvula.
4. Certhidea olivasea.

This image is by John Gould (14.Sep.1804 - 3.Feb.1881) - From "Voyage of the Beagle" as found on http://darwin-online.org.uk/converted/published/1845_Beagle_F14/1845_Beagle_F14_fig07.jpg]; also online through Biodiversity Heritage Library at https://www.biodiversitylibrary.org/page/2010582., Public Domain, https://commons.wikimedia.org/w/index.php?curid=3918303. This work is in the public domain in its country of origin and other countries and areas where the copyright term is the author's life plus 100 years or fewer. No changes were made to this image.

Darwin's "Sandwalk" at Down House in Kent, England, was his usual "thinking path"

This image is by Tedgrant at English Wikipedia - Transferred from en.wikipedia to Commons., Public Domain, https://commons.wikimedia.org/w/index.php?curid=3747372; en.wikipedia.org/wiki/Charles_Darwin#/media/File:Darwins_Thinking_Path.JPG. This work has been released into the public domain by its author, Tedgrant at English Wikipedia. No changes were made to this image.

Author's preface to Amazing Evolution Volume 2

Scientists have estimated that, to date,10% of all species on Earth have been identified. Thus, 90% have not yet been identified. The number of species wild creatures on Earth, excluding plants, is truly staggering. Estimates suggest there are around 8.7 million species on the planet, with approximately 6.5 million species on land and 2.2 million in the oceans. 8.7 million multiplied by 90 = 783 million species yet to be identified.

Researchers on an expedition to Peru in 2024 found four new mammal species, eight new fish, three new amphibians and ten new butterfly species. The Natural History Museum in London said its researchers had been involved in 190 new discoveries of living and fossilized animals, including 11 new species of moth, eight crabs, four rats and four snakes. Scientists linked to Kew Botanical Gardens, London, described 149 new species of plant and 23 new species of fungi.

Among the finds from Kew Gardens was an intriguing new species of fungi in wooded heathland near the town of Royal Tunbridge Wells, England: *Phellodon castaneoleucus* sports teeth-like structures instead of the gills usually seen beneath mushroom caps.

A species of Phellodon. This one is *Phellodon sinclairii*, a native tooth fungus found in beech forests in New Zealand.

Botanists also discovered five new orchid species from sites across the Indonesian archipelago, a gray-stemmed ghost palm from western Borneo with leaves with white undersides, and an enigmatic family of plants known as Afrothismia that are confined to continental African forests without the ability to photosynthesize.

Seems like there is some way to go and lots of work for keen young zoologists

Author's preface to Amazing Evolution Volume 3

How many books can be written about evolution? The number is the same as the number of species of wildlife, including plants, present on Earth, since all have evolved to a greater or lesser extent. There are over 100 million insects on the planet alone. This is my third book on the subject, so maybe a few more to come.
Here is an astonishing story to start Volume 4 with.

Amazing evolution volume 4

1. The wily Hedge sparrow

Hedge sparrows are ostensibly the most unassuming of birds. They like to skulk in garden hedges, although they do venture onto garden lawns to feed. They are not related to birds of the large Sparrow family. They do have some sense of adventure since, in countries with forested mountains, they migrate vertically in summer, returning to lower levels when (the northern hemisphere) summer is over.

Figure 1. Hedge sparrow (aka Dunnock or Hedge accentor) atop a tree in the mountains of Bulgaria during summer

However, it is their rapacious behaviour to ensure that their genes are passed on that is mind-boggling: male Dunnocks will grasp, and remove from a female Dunnock's cloaca, sperm that has been deposited there by

rival male. Is this "locking the stable door after the horse has bolted"? Probably, but it does seem a bit extreme. "A wolf in sheep' clothing at least".

(A cloaca or vent, is a rear orifice that serves as the only opening for the digestive, reproductive, and urinary tracts of many vertebrate animals and all amphibians, reptiles, birds, and cartilaginous fish).

2. Bird beaks

It was the bills (or beaks) of finches in the Galapagos Islands that gave Darwin his major Eureka moment: their bills were adapted to the flowering plants that surrounded them, thus giving them a survival advantage over the cousin relatives. Many other adaptations of bird bills have occurred, and it is all to do with finding food. Most birds have bills that are straight and fairly short - think European Blackbirds, thrushes and wagtails – and Dunnocks! Some are straight but long as in many wading (shorebirds) such as Oystercatchers or Grey-tailed tattlers.

Figure 2. Pied Oystercatchers in Australia

Figure 3. Grey-tailed Tattlers about to leave Australia's summer to breed in Siberia

Some have intermediate-sized bills such as terns, herons and kingfishers, which use them for catching fish.

Figures 4 - 5. Caspian tern – a global species

Figure 6. Mangrove heron from the southern hemisphere

Figure 7. A Nankeen night heron. (Nankeen is used in Australia to mean rusty)

Figures 8 - 9. Forest Kingfisher from Australia

Others are curved and the curve may be downwards as in the case of curlews, or upwards as in the case of avocets and the Terek sandpiper – or even sideways as in the case of the Wrybill, which is endemic to New Zealand

Figure 10. Eastern Curlew

Figure 11. Red-headed avocet

Figure 12. A Wrybill showing its beak bent to the right. It is the only species of bird in the World with a beak that is bent sideways in one direction, always to the right. They have evolved in a very distinct way to thrive on braided riverbeds: their beak is slightly tilted towards the right at an angle of about 12–26 degrees, which is

speculated to help them get insect larvae and small invertebrates that take refuge under the crevices of rocks. (A braided river consists of a network of river channels separated by small, often temporary, islands called braid bars).

This image is by Pseudopanax at English Wikipedia - Own work, Public Domain, https://commons.wikimedia.org/w/index.php?curid=174192075; https://en.wikipedia.org/wiki/Wrybill#/media/File:Side_view_of_wrybill_standing_on_the_sand,_looking,_showing_its_curved_beak.jpg. The copyright holder of this work has released this work into the public domain.

Some are tiny such as those of swallows, who really only need an orifice for catching small insects in flight.

Figure 13. Welcome swallow in Australia

One of the most unusual bills is that of the Black skimmer, a bird found in North and South America: the upper mandible is significantly shorter than the lower mandible. This has developed because of the bird's method of hunting. It flies low over the water with its longer lower bill slicing through the water. When it contacts a fish, the upper mandible slams shut, entrapping its prey.

Figure 14. Black skimmer

And then there are hooked bills – found in birds of prey, vultures, owls, parrots and cockatoos. Eagles, kites, harriers, falcons, hawks and vultures use them for tearing flesh off their prey. Parrots and cockatoos use them for obtaining their natural food from plants and eating the seeds.

Figure 15. Goden eagle

This image is by Zoo Ostrava - Zoologická zahrada Ostrava, Public Domain, https://commons.wikimedia.org/w/index.php?curid=2557459;

en.wikipedia.org/wiki/Golden_eagle#/media/File:Orel_skalní_2.jpg. This work has been released into the public domain by its author, Zoo Ostrava.

Figures 16. Peregrine falcon

This image is by USFWS photo - fuseaction=records.display&CFID=11899291&CFTOKEN=10439201&id=B553F510%2DE093%2D4388%2D97CE9CA9A2C51283, Public Domain, https://commons.wikimedia.org/w/index.php?curid=2670880; en.wikipedia.org/wiki/Peregrine_falcon#/media/File:Falco_peregrinus_nest_USFWS.jpg. This image or recording is the work of a U.S. Fish and Wildlife Service employee, taken or made as part of that person's official duties. As a work of the U.S. federal government, the image is in the public domain.

Figure 17. A pair of peregrines eating ducks

This image is by John James Audubon - Birds of America, Public Domain, https://commons.wikimedia.org/w/index.php?curid=437571; en.wikipedia.org/wiki/Peregrine_falcon#/media/File:Audubon-peregrinus.jpg. The author died in 1851, so this work is in the public domain in its country of origin and other countries and areas where the copyright term is the author's life plus 100 years or fewer.

Figure 18. African fish eagle

Figure 19. Spotted Eagle-Owl

Figure 20. Australian King Parrot

Figure 21. Red-collared Lorikeets drinking and bathing

Figure 22. Sulphur-crested cockatoo

3. IRIDESCENT FEATHERS

Iridescence is the phenomenon of certain surfaces that appear to change colour as the angle of viewing, or the angle of illumination, changes. Iridescence is caused by interference of light waves in microstructures or in thin films. Examples of iridescence in the natural world include feathers and butterfly wings.

Iridescent feathers are produced by nanostructures in feathers and have evolved in many bird species. Iridescent feathers get their colour from interference due to waves of light interacting with each other to produce either constructive reinforcement, or destructive cancellation of particular wavelengths, or colours. Examples are found in hummingbirds, peacocks, European starlings and some swallow species.

Many species of hummingbirds exhibit iridescence, especially of their throat feathers.

One example is the Bee hummingbird, which is endemic to Cuba

Figures 23 - 24. Bee hummingbird: when the sun is not shining directly on its throat feathers, the throat feathers appear black.

When the sun shines directly on its throat feathers these now appear red – due to iridescence.

Iridescence is thought to be used in communication, territorial displays, and attracting mates.

4. **BIRDS CAN SEE COLORS IN THE ULTRAVIOLET RANGE**

Did you know that many species of birds, unlike humans, can detect colors in the ultraviolet (UV) range? Such birds have dedicated cone cells in their retinas for perceiving wavelengths in the ultraviolet and violet regions of the light spectrum. The major bird families that have UV vision are Palaeognathae (ratites and tinamous), Charadriiformes (shorebirds, gulls, and alcids), Trogoniformes (trogons), Psittaciformes (parrots), and Passeriformes (perching birds), representing more than half of all avian species.

UV vision can be useful for courtship. Male blue tits have an ultraviolet reflective crown patch which is displayed in courtship by posturing and raising of their nape feathers[1].

The bill's appearance is important in the interactions of the European Blackbird. Although the UV component seems unimportant in interactions between territory-holding males, where the degree of orange is the main factor, the female responds more strongly to males with bills that have good UV-reflectiveness[2].

Insects, reptiles, and crustaceans are also sensitive to short wavelengths of light.

5. CAMOUFLAGE – SPOT THE CREATURE

Figure 25. A Slender-tailed Nightjar sitting on the ground

Figure 26. Striated (Mangrove) Heron: clue – in the middle of the photograph in between the wires

Figure 27. *Synanceia nana* – a species of stonefish. Which is the rock and which is the fish?

This image is by Photo2222 at English Wikipedia - Transferred from en.wikipedia to Commons., Public Domain, https://commons.wikimedia.org/w/index.php?curid=2356773; https://en.wikipedia.org/wiki/Synanceia#/media/File:Synanceia_nana.JPG. This work has been released into the public domain by its author, Photo2222 at English Wikipedia.

Figure 28. Tawny Frogmouths blend in with the colour and texture of tree bark

Tawny frogmouths are found throughout most of the Australian mainland. They can be found in almost any habitat type, including forests and woodlands, scrub and heathland vegetation, and savannahs.

They are often killed or injured on rural roads during feeding, as they fly in front of cars when chasing insects illuminated in the beam of the headlights[5].

This image is Copyrighted free use, https://commons.wikimedia.org/w/index.php?curid=540221; https://en.wikipedia.org/wiki/TawThe copyright holder of this work allows anyone to use it for any purpose including unrestricted redistribution, commercial use, and modification.ny_frogmouth#/media/File:Tawny_Frogmouth_(Coverdale).jpg.

Frogmouths are related to the potoos, one member of which is the Great Potoo: wha a wonderful name – straight out of Alice In Wonderland.

Indian Leaf Butterfly

When sitting on a tree trunk surrounded by dead leaves, the Indian leaf butterfly's resemblance to its surroundings is uncanny.

Figure 29. Indian leaf butterfly with wings folded

This image is by user:peellden - Own work, CC BY-SA 3.0, en.wikipedia.org/wiki/Kallima_inachus#/media/File:Kallima_inachis_formosana.jpghttps://commons.wikimedia.org/w/index.php?curid=1816953; Permission is granted to copy, distribute and/or modify this document under the terms of the GNU Free Documentation License, Version 1.2 or any later version published by the Free Software Foundation.

But this appearance is probably not that attractive to a member of the opposite sex. So what does it do? It opens its wings to reveal a panoply of blue and yellow.

Figure 30. Indian leaf butterfly with wings open

6. HOW DO CHAMELEONS CHANGE THE COLOR OF THEIR SKIN?

Chameleons have two superimposed layers within their skin that control their colour and thermoregulation. The top layer contains a lattice of guanine nanocrystals, and by exciting this lattice the spacing between the nanocrystals can be manipulated, which in turn affects which wavelengths of light are reflected and which are absorbed. Exciting the lattice increases the distance between the nanocrystals, and the skin reflects longer wavelengths of light. Thus, in a relaxed state the crystals reflect blue and green, but in an excited state the longer wavelengths such as yellow, orange, green, and red are reflected.[3]

7. HOW DO ANIMALS MAKE THEIR WEIRD AND WONDERFUL SOUNDS?

In the animal world making sound is key to survival. Cicadas do it, lyrebirds do it, and even fish in the ocean do it. Male animals are especially vocal when they are trying to attract a mate. Sound is created by vibration. When an object vibrates it causes the air molecules around it to vibrate. This in turn creates ripples that hit our eardrums causing them to vibrate and produce a signal that is detected by the brain and interpreted as sound. Lots of animals make sound by using air from their lungs to make their vocal cords. It is well known that elephants make a trumpeting sound and that lions and tigers roar, but what about smaller creatures… like fish?

Fish

Even fish, that have nothing remotely resembling the anatomical features of mammals, can make sound.

How do they do it?

Some grind their bones together, while herrings are believed to communicate by blowing bubbles out their anus. Australian fish have generally been found to make sounds by using special 'sonic' or 'drumming' muscles to 'strum' their swim bladders - balloon-like organs that they use to help keep them buoyant.

Some fish are soloists: a hooting Batfish, a knocking Catfish, a beeping Terapon, and a honking Mullaway.

Batfish include

California batfish or bat ray (*Myliobatis californicus*), an eagle ray from the East Pacific

Red lipped batfish or the Galapagos batfish

Ogcocephalidae, a family of anglerfish found in oceans worldwide

Platax, a spadefish genus from the Indian and Pacific Oceans

Freshwater batfish (*Myxocyprinus asiaticus*), better known as the Chinese high fin banded shark

Members of the genus Halieutaea

Figure 31. A California batfish swimming

Figure 32. California batfish adopting a feeding posture

This image is by Tony Chess - http://swfsc.noaa.gov/textblock.aspx?ParentMenuId=123&id=986#, Public Domain, https://commons.wikimedia.org/w/index.php?curid=518338; https://en.wikipedia.org/wiki/Bat_ray#/media/File:Bat-ray-NOAA-Tony-Chess.jpg. This image is in the public domain because it contains materials that originally came from the U.S. National Oceanic and Atmospheric Administration, taken or made as part of an employee's official duties.

Figure 33. *Terapon jarbua*

This image is by Unknown author - Bloch, Marcus Elieser (1793) Naturgeschichte der ausländischen Fische. Siebenter Theil. Berlin: J. Morino & Comp. Taf. CCXXXVIII. Fig. 1, Public Domain, https://commons.wikimedia.org/w/index.php?curid=114913493;

en.wikipedia.org/wiki/Terapon_jarbua#/media/File:Therapon_servus.png. This work is in the public domain in its country of origin and other countries and areas where the copyright term is the author's life plus 100 years or fewer. PD 1929.

Fish also "sing" in choruses", including the noise made by a plethora of plankton-eating fish on the Great Barrier Reef.

Mammals

The Koala

The male Koala is renowned for his loud vocal displays designed to impress females and announce his territory. First, he starts off making repeated snore-like snorts as he inhales, and then makes a grunty belch sound as he exhales. As he picks up speed the sounds turn into an incredibly loud bellowing.

The larger the male, the lower his sound, and the further it travels to advertise his size and prowess. Apparently, male koala sounds are highly individual and help them be identified. Female koalas tend to screech and cry rather than bellow, and usually only if they are distressed or being pursued by an unwanted male.

Figure 34. US President Barack Obama with a Koala

Bats

Small bats like the Large-footed myotis (*Myotis macropus*) emit ultrasound noises. The frequencies are above 20,000 Hertz and beyond the ability of humans to hear. The bats produce these waves for navigation and finding prey (echolocation). Not that they're blind: they just need a way to sense their environment when they hunt at night. The pitch of a sound (whether it's high or low) is the rate of vibrations per second - known as frequency – and measured in Hertz (Hz).The Large-footed myotis trawls for fish and insects by flying just 15 centimetres above the surface of water in the dark, so it has to know exactly when to reach out its claws to grab its prey.

To do this, the bat produces pulses of ultrasound that bounce back off the water surface and gives it information on what is there. Once it picks up something promising it sends out more pulses to get a clearer picture.

Figure 35. Sonograms of two species of bat: comparison of *Myotis macropus* call (left) and *Nyctophilus gouldi* call (right). Those of *Myotis Macropus* are just visible as vertical lines, while those of *Nyctophilus gouldi* are more easily seen.

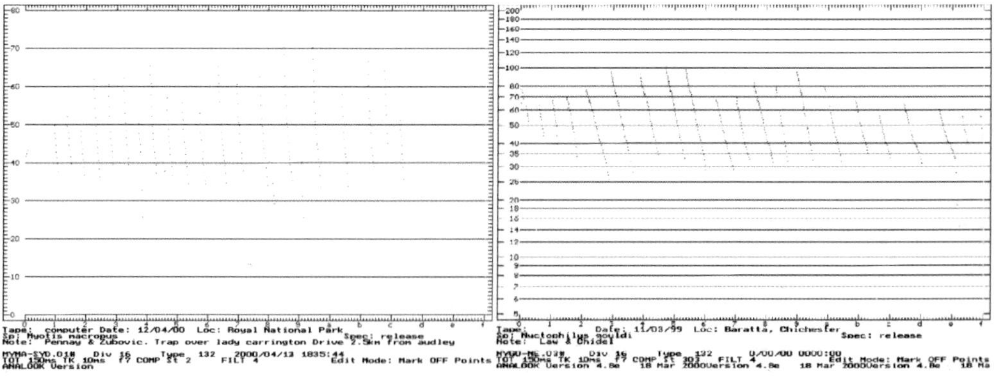

Blue whales

At the opposite end of sound frequencies that humans cannot hear is 'infrasound' — below 20Hz. These are the frequencies exploited by one of the largest and loudest animal in the world: the Blue whale.

Blue whales sing by vibrating a U-shaped fold of skin that sits between their lungs and a special air sac, which also helps amplify the sound, producing resonance. By circulating the air back and forth they are able to make continuous calls.

The low-frequency soundwaves that Pygmy blue whales produce can be heard hundreds of kilometres away. Males sing to attract females especially when they are moving north heading towards their breeding grounds.

Amphibians

Frogs

Male frogs usually call all night. Frogs do not have ribs and a diaphragm, so they have to go to a lot of effort to fill their lungs with air that they can then use to vibrate their vocal folds. First, they take air in through their nostrils by lowering the floor of their mouth (to create a negative pressure). Then they block their nostrils and lift the floor of their mouth to push the air into their lungs. They do this repeatedly until their lungs are full. In some frogs this is aided by a vocal sac - which can balloon out below their chin, or in some cases to either side - to recycle the air so they can keep making noise without gasping for air. The sac also acts as a resonator to help distribute the calls far and wide.

Figure 36. A Bleating tree frog showing its vocal sac underneath its chin

Birds

Now for one of the great bird singers - the Lyrebird

The Superb lyrebird has to be the cleverest songbird of all. Apart from a range of whistles and cackling that they use for their own song, it is famous for its mimicry. Like other birds, the secret to the songs it produces lies in an organ called the syrinx, which is located where the windpipe branches into two in the chest.

The lyrebird's super flexibility in muscles around this organ means they can rapidly and independently control airflow through the two branches of their windpipe, producing all kinds of sounds in rapid succession.

Lyrebirds do not just mimic other birds. If they are bred in captivity, they also mimic sounds from the human world – from chainsaws and construction sounds to evacuation alarms and even human voices. My favorite is their imitation of a camera clicking

Insects

The prize for the loudest insect in the world goes to the cicada with some large species reaching more than 100 decibels – which is somewhere between the volume of a blender and a baby crying.

Australia is the cicada capital of the World, with 800 known species.

As male cicadas start up their calls to attract a mate or deter predators, others join in and together they create a deafening chorus.

Like frogs and birds, every species of cicada has its own unique sound, which is how females manage to recognise the right males.

The loudness of a sound is determined by the size of the sound waves - their amplitude, as measured in decibels (dB). At the heart of their sound-generating machinery are membranes in the abdomen called timbals. Like a tin clicker, they make a repetitive clicking sound by flexing a series of convex ribs along these membranes – buckling and straightening them several hundred times a second. The sound is then amplified in an air sac in the cicada's abdomen, before being pushed out through holes in its underside.

Cicadas are the loudest insects in the world.

Figure 37. The periodical cicada (*Magicicada septendecim*)

8. THE NAKED MOLE-RAT

It is not a pretty animal but it is an extraordinary animal. The Naked mole-rat lives in East Africa and spends its life underground burrowing in sandy soils[6]. It has an almost entirely cold-blooded form of body temperature regulation, as well as exhibiting, a complex social structure that includes a reproductive division of labor, separation of reproductive and non-reproductive castes, and cooperative care of young[7]. It lacks pain sensitivity in their skin, and has very low metabolic and respiratory rates. The animal also is remarkable for its longevity with lifespans in excess of 37 years[8].

Figure 38. A Naked mole-rat eating

This image is by Ltshears - Trisha M Shears - Own work, Public Domain, https://commons.wikimedia.org/w/index.php?curid=1732225; https://upload.wikimedia.org/wikipedia/commons/c/cf/Naked_Mole_Rat_Eating.jpg. I, the copyright holder of this work, release this work into the public domain.

They are well adapted to their underground existence. Their eyes are quite small, and their visual acuity is poor. Their legs are thin and short; however, they are highly adept at moving underground and can move backward as fast as they can move forward. Their large, protruding teeth are used to dig and their lips are sealed just behind the teeth, preventing soil from filling their mouths while digging. They have little hair and wrinkled pink or yellowish skin. Only one female (the queen) and one to three males reproduce, while the rest of the members of the colony function as workers. In colonies (usually of 75 – 80 individuals), the dominant species is female: the queen.

9. WOODPECKERS' SPECIAL FEATURES

Woodpeckers have four toes, the first and the fourth facing backward and the second and third facing forward. (The Three-toed woodpecker of North America is an anomaly, but I will come back to that). Woodpeckers climb up the trunks of trees in short hops but during these hops their tail feathers, known as ratrices, are pressed against the tree trunk to give them upward stability while climbing. Woodpeckers also have very strong neck muscles to enable them to peck for long periods at a time when carving out a nesting hole. Do woodpeckers get headaches from their drilling activity?

Members of this family are found worldwide, except for Australia, New Guinea, New Zealand, Madagascar. Since they are present in Indonesia, it is unclear why they are not found in Australia – a fairly short distance from Indonesia (which has woodpeckers). Likewise, Madagascar is not that far from Africa, which has woodpeckers. One theory about Australia is that the majority of trees are eucalypts and the wood is hard, so maybe they *would* get headaches from constant drilling.

Figure 39. A Green woodpecker from Europe showing the use of its tail feathers while climbing up a tree trunk

Downloaded from Unsplash. Photo credit Phil Baum

Figure 40. A Pileated woodpecker from North America using its ratrices

Downloaded from Unsplash. Photo credit Zdeněk Macháček

The American Three-toed Woodpecker has three toes. It has been suggested this is due to the evolutionary loss of the fourth toe - an unusual trait shared only by the Eurasian Three-toed and Black-backed Woodpeckers. With only three toes, these species may be able to lean farther away from the tree and thereby hit the tree harder than other woodpeckers, all of which have four toes.

A typical woodpecker will bang its head against wood 20 times per second, accelerating 1,200 times more than gravity. In an average day, a woodpecker does this around 12,000 times. To prevent brain damage from the rapid and repeated powerful impacts, woodpeckers have a number of physical features that protect their brains[9]. These include a relatively small and smooth brain, narrow subdural space, little cerebrospinal fluid surrounding it to prevent it from moving back and forth inside the skull during pecking, the orientation of the brain within the skull (which maximises the contact area between the brain and the skull) and the short duration of contact. The skull consists of strong but compressible, sponge-like bone, which is most concentrated in the forehead and the back of the skull[9]. Another anatomical adaptation of woodpeckers is the enormously elongated hyoid bone which subdivides, passes on either side of the spinal column and wraps around the brain case, before ending in the right nostril cavity. It plays the role of safety-belt[10].

Computer simulations have shown that 99.7% of the energy generated in pecking is stored in the form of strain energy, which is distributed throughout the bird's body, with only a small remaining fraction of the energy going into the brain. The pecking also causes the woodpecker's skull to heat up, which is part of the reason why they often peck in short bursts with brief breaks in between, giving the head some time to cool. During the millisecond before contact with wood, a thickened nictitating membrane closes, protecting the eye from flying debris[11]. These membranes also prevent the retina from tearing.

10. FISHY TALES

Gulper eel

The Gulper (aka Pelican) eel’s defining trait is its expandable mouth, dwarfing its body when fully extended. This adaptation allows it to consume a wide range of prey. It also has a glowing tail tip to lure its quarry. It has been found in the temperate and tropical areas of all oceans. In the North Atlantic, it seems to have a range in depth from 500 to 3,000 m (1,600 to 9,800 ft).

Figure 41. Gulper eel

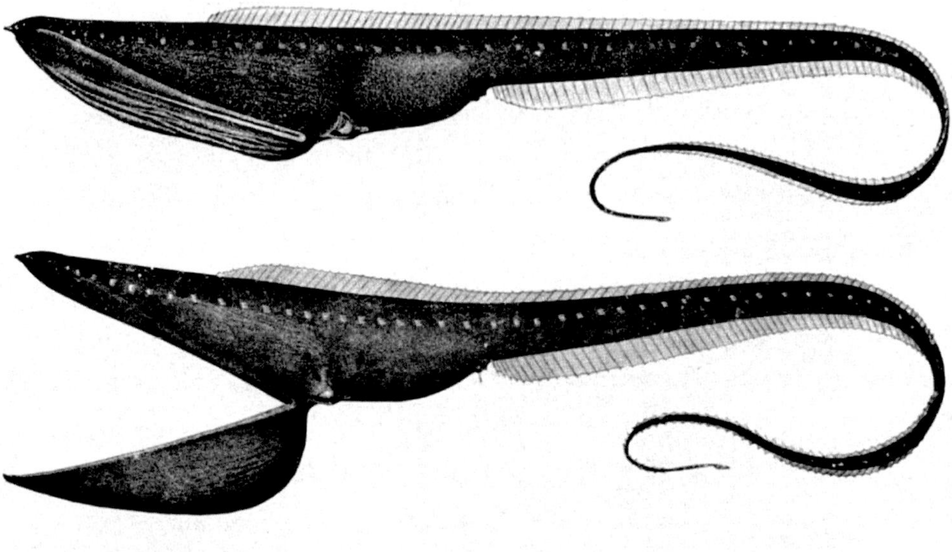

This image is by Public Domain, https://commons.wikimedia.org/w/index.php?curid=634435; en.wikipedia.org/wiki/Pelican_eel#/media/File:Eurypharynx_pelecanoides.jpg. From plate 49 of Oceanic Ichthyology by G. Brown Goode and Tarleton H. Bean., published 1896. This work is in the public domain in its country of origin and other countries and areas where the copyright term is the author's life plus 70 years or fewer.

Figure 42. X-ray image of the Gulper eel

This image is by Sandra Raredon/Smithsonian Institution - cropped version of File:Eurypharynx pelecanoides X-ray.jpg, Public Domain https://commons.wikimedia.org/w/index.php?curid=56065550; https://en.wikipedia.org/wiki/Pelican_eel#/media/File:Eurypharynx_pelecanoides_X-ray_(cropped).jpgn. This work is in the public domain in the United States because it is a work prepared by an officer or employee of the United States Government as part of that person's official duties under the terms of Title 17, Chapter 1, Section 105 of the US Code.

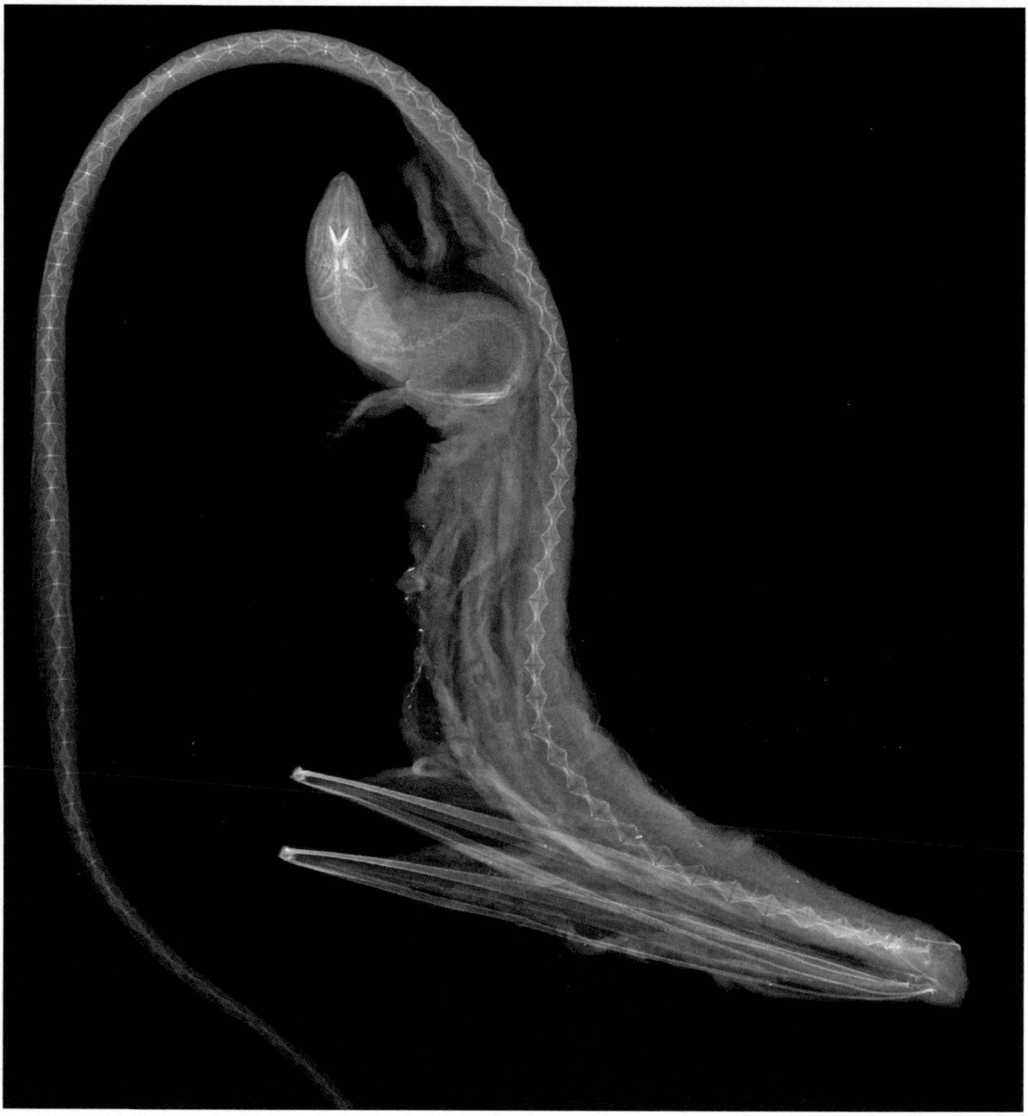

Giant squid

Recent deep-sea footage confirmed the Giant squid's remarkable anatomy with eyes the size of dinner plates and feeding tentacles 10 m (33 ft) in length for capturing prey[12].

Figure 43. A giant squid specimen measuring over 4 m (13 ft) without its two long feeding tentacles.

This image is by Public Domain, https://commons.wikimedia.org/w/index.php?curid=33879; https://en.wikipedia.org/wiki/Giant_squid#/media/File:Giant_Squid_NASA.jpg. This file is in the public domain in the United States because it was solely created by NASA.

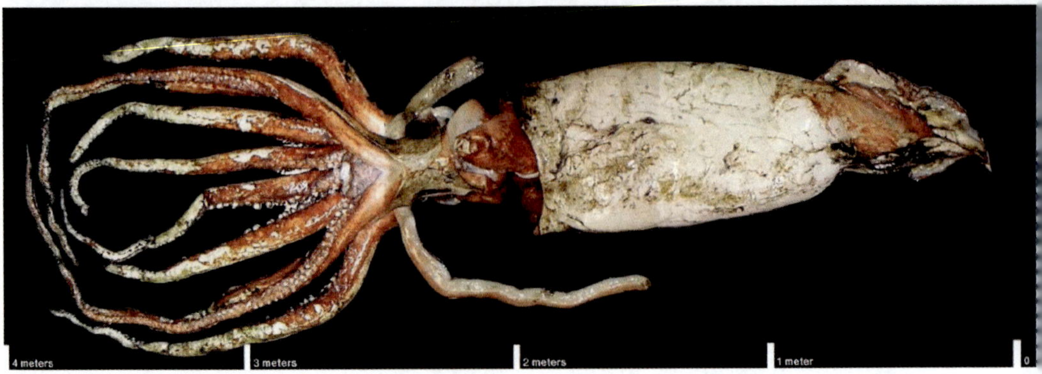

Figure 44. Global distribution of the Giant squid

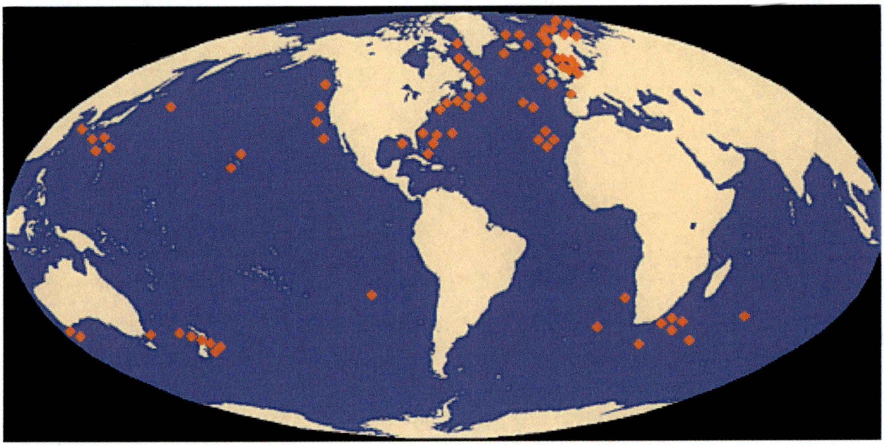

This image is by Public Domain, https://commons.wikimedia.org/w/index.php?curid=34809833; https://en.wikipedia.org/wiki/Giant_squid#/media/File:Architeuthis_distribution.png. This file is in the public domain in the United States because it was solely created by NASA.

Frilled shark

The name, Frilled shark, derives from the fringed appearance of the six pairs of gill slits at the shark's throat. It is considered a living fossil, because of its primitive, eel-like physical traits, such as its dark-brown color, the articulation of the jaws to the cranium, and a 2.0 m (6.6 ft) long body, which has dorsal, pelvic, and anal fins located towards the tail. Frilled sharks are distributed throughout regions of the Atlantic and the Pacific oceans, usually in the waters of the outer continental shelf and of the upper continental slope, where they live close to the ocean floor, near biologically productive areas of the ecosystem. When hunting food, it curls its tail against a rock and moves like an eel, bending and lunging to capture and swallow whole prey with its long and flexible jaws, which are equipped with 300 recurved (inward pointing) needle-like teeth[13].

To live on a diet of cephalopods, smaller sharks, and bony fish, the Frilled shark practices nocturnal vertical migration to feed at the surface of the ocean.

Figure 45. The Frilled shark (*Chlamydoselachus anguineus)*

This image is by Robert Mintern - Report on the deep-sea fishes collected by H.M.S. Challenger during the years 1873-1876, Plate LXIV. Günther, Albert C. L. G. (Albert Carl Ludwig Gotthilf), 1830-1914, Public Domain, https://commons.wikimedia.org/w/index.php?curid=10804316; https://en.wikipedia.org/wiki/Frilled_shark#/media/File:Chlamydoselachus_anguineus1.jpg. The author died in 1908, so this work is in the public domain in its country of origin and other countries and areas where the copyright term is the author's life plus 100 years or fewer. This work is in the public domain in the United States because it was published (or registered with the U.S. Copyright Office) before January 1, 1930.

The Olive perchlet

The Olive perchlet is a five to six centimetre fish that plays a key role in its ecosystem by eating small water insects. The species was thought to have become extinct in the Australian states of Victoria and South Australia, while populations in New South Wales and Queensland survived. A breeding program in Victoria, however, produced 200 healthy adults, which were released in a northern national park in February 2023. They live in the vegetated edges of lakes, creeks, swamps, wetlands and rivers, usually in areas with little or no flow, particularly backwaters.

Releasing the perchlet is part of a long-term plan to repopulate species on restored flood plain habitats throughout the Murray River corridor.

Figure 46. The Olive perchlet

Murray-Darling Basin Authority. Photo credit Warren Chad

Xingren golden-line fish

The Xingren golden-line fish (*Sinocyclocheilus xingrenensis*), lacks scales and has unusually large eyes, suggesting it is still adapting to its underground habitat. However, the Xingren golden-line fish still has well-developed eyes, meaning it likely experiences some exposure to light. Research suggests that cave windows - gaps in the rock that allow daylight to filter through - may be influencing its evolution, preventing total blindness from developing.

The Xingren golden-line fish also stands out because it lacks a horn-like structure found in some of its closest relatives, such as *S. longicornus*, a species discovered in 2023 with a unicorn-like protrusion on its head. These subtle differences show how cave fish evolve along different paths, even within the same genus.

12. STRANGE DENIZENS OF BASS CANYON

Four thousand metres below the surface of the Tasman Sea lies a mysterious and alien world. Bass Canyon is an 80-kilometre-long gash in the sea floor - a place of near complete darkness and crushing pressures. The temperature in the canyon is barely above freezing, and the conditions are seemingly incompatible for life. But the Bass Canyon is far from lifeless. There is a staggering diversity of strange creatures, the likes of which rarely see the light of day.

The Australian Commonwealth Scientific and Industrial Research Organisation's (CSIRO) Research Vessel Investigator explored this submarine crevice with a combination of multi-beam sonar and visual camera photography during an autumn trip off the Tasmanian coast.

Here are some of their findings.

Tripod fish

Figure 47. Tripod fish

This fish walks on the ocean floor using three elongated fins

This image is by Bahamas Deep-Sea Coral Expedition Science Party, NOAA-OER. - http://www.photolib.noaa.gov/htmls/expl1868.htm, Public Domain, https://commons.wikimedia.org/w/index.php?curid=16328440; en.wikipedia.org/wiki/Bathypterois_grallator#/media/File:Tripod_fish.jpg. This image is in the ***public domain*** because it contains materials that originally came from the U.S. *National Oceanic and Atmospheric Administration*, taken or made as part of an employee's official duties.

Figure 48. Another species of Tripod fish

This image is by NOAA Okeanos Explorer Program, INDEX-SATAL 2010, NOAA/OER - http://www.photolib.noaa.gov/htmls/expl2148.htm, Public Domain, https://commons.wikimedia.org/w/index.php?curid=16328994; https://en.wikipedia.org/wiki/Aulopiformes#/media/File:Tripod_fish1.jpg. This image is in the public domain because it contains materials that originally came from the U.S. National Oceanic and Atmospheric Administration, taken or made as part of an employee's official duties.

Lizardfish

Lizardfish (*Bathysaurus mollis*)

Lizardfish is a deep-sea ambush predator with a large mouth filled with needle-sharp depressible teeth. It has so many rows of teeth that they remain visible even when the lizardfish closes its mouth. There are several species of lizardfish.

Figure 49. Lizardfish (*Bathysaurus mollis*)

This image is by NOAA/MBARI - http://www.mbnms-simon.org/other/photos/photo_info.php?photoID=182, Public Domain, https://commons.wikimedia.org/w/index.php?curid=1935640;

en.wikipedia.org/wiki/Bathysaurus_mollis#/media/File:Bathysaurus_mollis.jpg. This image is in the public domain because it contains materials that originally came from the U.S. National Oceanic and Atmospheric Administration, taken or made as part of an employee's official duties.

Other lizardfish

Other species of lizardfish live in shallow coastal waters. They all have mouths full of sharp teeth, even on the tongue[13].

Figure 50. Diamond lizardfish (*Synodus synodus*)

This image is by Paul Louis Oudart (1796–1860)- Voyage dans l'Amérique méridionale, Public Domain, https://commons.wikimedia.org/w/index.php?curid=18650490; https://en.wikipedia.org/wiki/Diamond_lizardfish#/media/File:Synodus_synodus_Orbigny.jpg. This work is in the public domain in its country of origin and other countries and areas where the copyright term is the author's life plus 100 years or fewer.

Figure 51. The Inshore lizardfish (*Sydodus foetens*)

This image is by NOAA AOML - https://www.aoml.noaa.gov/news/juvenile-sportfish-monitoring/, Public Domain, https://commons.wikimedia.org/w/index.php?curid=99362178; https://upload.wikimedia.org/wikipedia/commons/9/93/Sportfish8.jpg. This image is in the public domain because it contains materials that originally came from the U.S. National Oceanic and Atmospheric Administration, taken or made as part of an employee's official duties.

13. BIRDS' SENSE OF SMELL

Birds do not typically have a strong sense of smell. A measure of an animal's sense of smell is the size of its olfactory bulbs. The olfactory bulb is a neural structure of the vertebrate forebrain involved in olfaction, the sense of smell. Larger olfactory bulbs indicate a more developed sense of smell.

Figure 52. The human olfactory bulb. The olfactory bulb is composed of the two parallel structures near the front of the brain marked (F) in this image

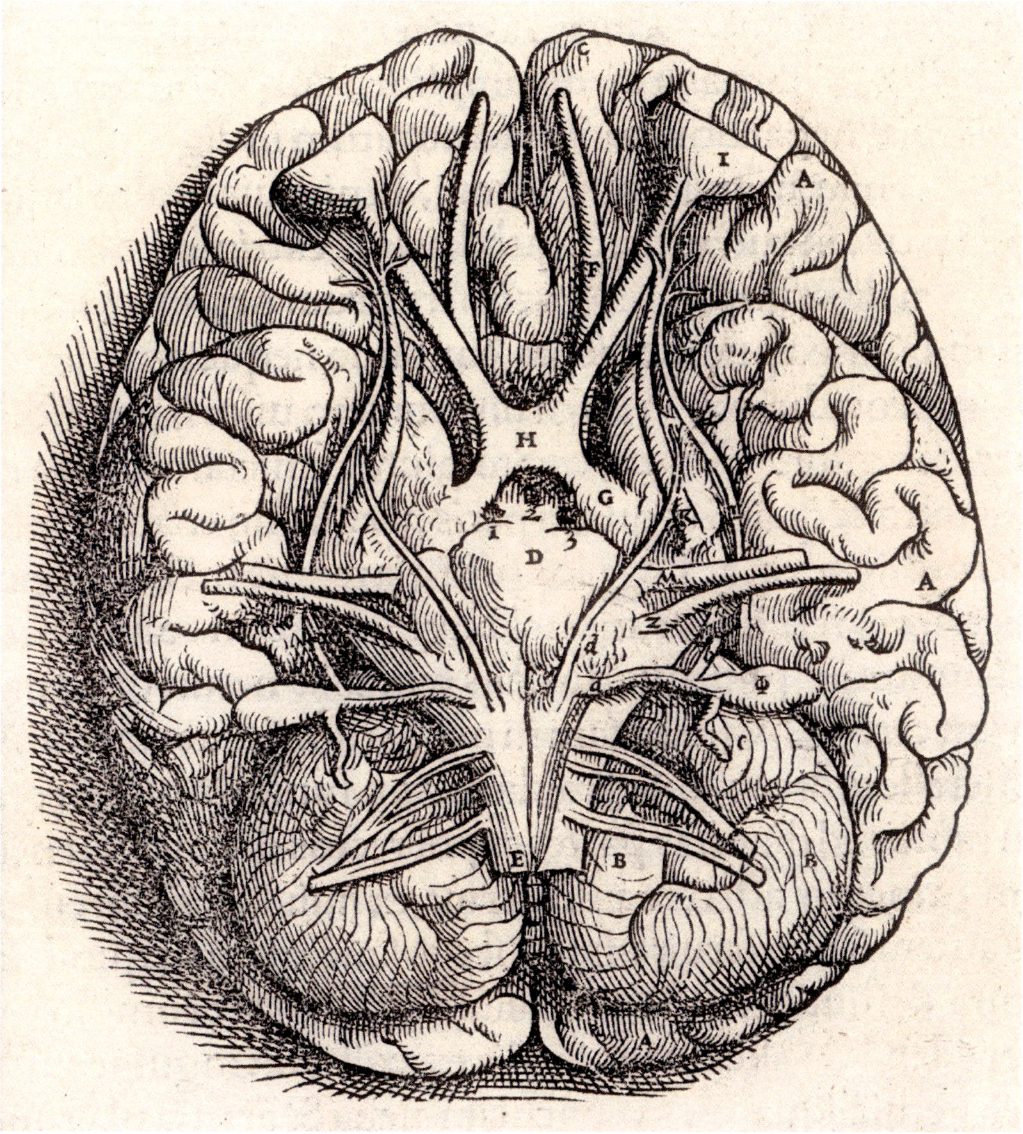

Turkey Vultures have a powerful sense of smell and they have one of the largest olfactory systems of all birds. They can smell carrion (the decaying flesh of dead animals) from over a mile away.

The Turkey vulture's anatomical and physiological adaptations are unparalleled by any other animal. Their large and well-developed olfactory bulbs and powerful cellular mechanisms allow them to detect diluted volatile gas molecules in the atmosphere hundreds of feet away while flying.

Others birds use smell to find specific food sources, such as storm petrels locating krill.

Figure 53. Turkey vulture

This image is by John James Audubon and is in the public domain in those countries who have a copyright term of 100 years or fewer after the death of the author.

Figure 54. Black-bellied storm petrel

14. BIRDS' SENSE OF THE PRESENCE OF WATER

I feed the birds in our garden if it has been a cold wet night. On such an occasion they are gathered round the door to our deck, waiting. There are Sulphur-crested cockaoos, Rainbow lorikeets, Noisy miners, Pies currawongs and Australian ravens – the usual suspects. On one occasion, the night had been mild, but the birds were there on the deck. There was no sign of rain. Two hour later it began to rain hard. Had the birds sensed this? Were they able to detect an excess of water molecules in the air?

A similar thing happens when Lake Eyre in South Australia fills with water due to heavy rains in northern Australia. This happens very infrequently, but when it does this body of water is a magnet for waterbirds - ducks, pelicans, cormorants fly there.

Some of these species are normally found several hundred miles from Lake Eyre. So how do they know that the lake is filling up? It has been said that some species of birds have a strong sense of smell that they can use to find water sources. But what does rainwater smell like? The ability to detect an excess of water molecules in the alr seems more likely to me.

Other possible mechanisms
Atmospheric cues: birds may detect changes in humidity, air pressure, or scent carried by prevailing winds from the lake. These subtle environmental signals could act as long-distance alerts.

Auditory or visual signals: though less likely over vast distances, some researchers speculate that birds might respond to distant thunder or cloud formations associated with flooding rains.

Social signalling: birds that are closer to the lake may begin migrating and trigger a chain reaction - others follow based on flock behavior or learned patterns.
Innate timing and memory: Some species may have evolved to recognize seasonal rainfall patterns in the Lake Eyre Basin. They might "know" from past experience that water is likely present after heavy rains in Queensland or the Channel Country.

Satellite tracking insights: recent studies using GPS trackers show that birds can travel from as far as China, Russia, and Antarctica to reach Lake Eyre when it floods. This suggests that they are responding to environmental triggers rather than fixed migratory schedules.

Bottom line: we do not really know at present.

15. WHALE SONGS AND HUMAN SPEECH FOLLOW THE SAME PATTERNS

Communication is a fundamental aspect of many species, but human language has long been considered unique in its complexity and structure.

However, new research suggests that whales' songs may share striking similarities with human speech patterns, raising questions about how language evolved.

Hellen Garland, a researcher at the University of St Andrews in Scotland, has found that whales' songs may share striking similarities with human speech patterns. A hidden, language-like structure in whale songs was unexpected and emphasized the need for further research to understand whether these structures serve a comparable function for whales as they do for humans. This discovery broadens the scope of linguistic research, suggesting that patterns in communication might not be as uniquely human as previously thought. Scientists are now considering the possibility that language's evolutionary roots may be more deeply intertwined with nature than earlier theories suggested.

16. NARWHALS MAY USE THEIR TUSKS TO PLAY

Narwhals exhibit at least 17 recorded and identified behaviors when interacting with prey. Scientists say they have the first recorded video evidence of narwhals using tusks, not only to strike and manipulate fish during feeding, but also engage in what appears to be playful behavior.

Figure 55. Narwhal

This image is by NOAA United States. National Marine Fisheries Service - Cetaceans of the Channel Islands National Marine Sanctuary, Public Domain, https://commons.wikimedia.org/w/index.php?curid=45804576; https://en.wikipedia.org/wiki/Narwhal#/media/File:Monodon_monoceros_NOAA.jpg. This image is in the public domain because it contains materials that originally came from the U.S. National Oceanic and Atmospheric Administration, taken or made as part of an employee's official duties.

Scientists have observed few interactions of Narwhals in their natural habitat, creating speculation about the purpose of the species' distinctive spiralling tusk. Found predominantly in males, the tusk grows up to 10 feet (3 meters) long, and previous research has suggested it serves as a competitive display to secure mates. But now, with the help of drones, research conducted in the Canadian High Arctic has uncovered that a narwhal may use its tusk for more than just courtship.

Researchers identified and described 17 distinct behaviors of Narwhals involving prey. The findings revealed a wide range of interactions and dynamics between narwhals and fish as well as the extraordinary agility, precision and speed of their tusks to track moving targets[14]

One of the researchers Dr. Gregory O'Corry-Crowe said "that the fact that these animals are not actually hunting the fish, but exploring, manipulating and interacting with it. It was really a game changer."

"It is important to document Narwhals' unknown behaviors to better understand how these animals adapt in a rapidly shifting habitat as oceans warm and sea ice melts".

As the researchers analyzed their drone video footage, they noticed subtle nuances in the Narwhals' behavior. In one instance a narwhal interacted with a fish by repeatedly nudging it with its tusk without attempting to eat it.

When researchers observed a lack of aggression in some interactions between narwhals and fish, they realized these scenarios were more similar to a cat-and-mouse game, in which the animals were chasing or "playing" rather than hunting.

"Play is a very important aspect of developing behavioral skills that will be important in terms of survival in other ways, but the fact that they are engaged in this kind of work reminds us that these species have complex and interesting lives." Younger Narwhals also tended to imitate the behavior of older ones, indicating that social learning could influence how narwhals use their tusks.

O'Corry-Crowe emphasized the crucial role of collaboration with indigenous communities in the region, whose partnership and knowledge contributed to the success of the research.

While the study mainly focused on tusk use, researchers also discovered other important behavioral shifts, particularly in Narwhals' dietary habits.

They found an unexpected predator-prey competition dynamic. Glaucous gulls frequently attempted to steal fish from Narwhals, significantly reducing the amount of prey they were able to catch.

Although gulls are known for stealing food gathered by land animals, there are few observations of seabirds stealing food from marine animals. In response to these challenges, Narwhals may have to dive deeper into the ocean to hunt for larger or denser schools of fish. Narwhals usually feed on deep-sea animals, such as squid or halibut. However, researchers observed Narwhals hunting Arctic Char closer to the ocean's surface, which is a shift in their hunting behavior pattern. Narwhals are highly sensitive and vulnerable to the effects of climate change because of their close association with sea ice for food, navigation and refuge. The habitat range of the Narwhal is changing and will be more restricted as oceans warm.

17. WHAT MAKES TURTLES 'DANCE'

Turtles can detect all Earth-strength magnetic fields, ranging from around 25,000 nanoteslas to 65,000 nanoteslas. A tesla is a measure of magnetic field intensity.

To understand the magnetic receptors of turtles, researchers collected 14 to 16 newly hatched Loggerhead turtles each August from 2017 to 2020. The turtles emerged from eight to 10 different nests from Bald Head Island, North Carolina. The team housed the turtles in individual tanks with controlled water temperatures and a standard diet to simulate natural seawater conditions.

Figure 56. Loggerhead turtle hatchlings

This image is by David L. Jones - Author, Public Domain, https://commons.wikimedia.org/w/index.php?curid=5965739; https://en.wikipedia.org/wiki/Loggerhead_sea_turtle#/media/File:BabyLoggerheadTurtle1.jpg. The copyright holder of this work has released this work into the public domain.

Previous turtle experiments have used magnetic intensities with at least a 2,000-nanotesla field difference, but in this study the research team chose locations along the US East Coast and developed a coil system to produce fields between 2,000 and 10,000 nanoteslas.

Over a two-month conditioning period, the study team placed the turtles in small buckets of artificial seawater and exposed them to two different magnetic fields for equal durations. One field matched the magnetic force of a site in the Gulf of Mexico and was associated with food (the "rewarded" field), while the other simulated the magnetic flux of a site near New Hampshire and had no food (the "unrewarded" field).

Once the conditioning ended, turtles were tested again in both magnetic fields, but this time, neither contained food, allowing researchers to determine whether the turtles had learned to associate the "rewarded" field with feeding. In the "rewarded" field, all the marine reptiles exhibited some extent of “turtle dance” behavior, which included tilting their bodies vertically, holding their heads near or above the water’s surface, opening their mouths, quickly moving their front flippers, and sometimes even spinning in place, according to the study.

To confirm the consistency of these findings across different sites, the researchers conducted the same experiment using magnetic fields that mimicked those off the coasts of Cuba versus Delaware, Maine versus Florida, and two additional locations.

In each of the five trials, about 80% of the turtles showed more “dancing” in the “rewarded” fields compared with the “unrewarded,” demonstrating that this skill is used globally, not just in one specific location.

While the “turtle dance” is particularly charming, the researchers noted that this behavior may only occurs in captivity. However, the movement pattern provided a useful measure to show whether the turtles learned the magnetic field and correlated it to food.

After the initial experiment, the scientists tested 16 turtles again four months later to evaluate their long-term memory. Even without additional reinforcement, 80% of the Loggerheads showed greater “dancing” in the “rewarded” field, although the overall amount of movement was lower.

Turtles may remember magnetic conditioning for a much longer duration, since most Loggerheads leave their nesting beach as hatchlings and return around 20 years later to lay their first eggs.

A similar study was carried out by Dr. Kayla Goforth and her team. They added markings to the turtles' shells using colored nail polish to differentiate the experimental groups.

Once the researchers established that turtles respond to magnetic fields associated with food, they wanted to determine whether turtles utilize the same or different biological systems for their magnetic map (knowing where they are) and their magnetic compass (knowing which direction to go).

Using radiofrequency waves, a type of energy that can disrupt biological sensors, researchers tested whether the turtles could still detect their magnetic map and magnetic compass.

One group of turtles was tested without radiofrequency waves, while the other was tested with radiofrequency waves. Normally, turtles swim in a certain direction depending on the radiofrequency field they live in to stay within the right ocean currents for migration. However, when radiofrequency waves were present, the turtles swam randomly, indicating their compass was disrupted.

Their ability to recognize the magnetic map (or food-associated locations) remained intact, however, even with radiofrequency interference.

This understanding provided additional information about how sea turtles, and other animals, are able to navigate hundreds or thousands of miles across oceans that do not have obvious physical features to help with navigation.

To investigate further how turtles interpret magnetic cues, the study team examined the two key features of Earth's magnetic field: inclination, or the tilt of magnetic field lines relative to Earth's surface, and intensity, or the strength of the magnetic field.

The researchers generated mismatched magnetic fields by combining the inclination from one geographic location with the intensity from another, swapping the values throughout the trials.

Turtles did not recognize a place unless both inclination and intensity matched, indicating that they rely on a combination of these factors to determine their location. This research reveals that, as with birds and amphibians, turtles also rely on dual magnetoreception systems.

The most important conservation takeaway from this research is that radiofrequency waves produced by electronic devices negatively affect sea turtles' navigation.

If turtles reside in ocean areas with heavy boat traffic or come to beaches to nest where people use phones frequently, their navigational senses could be disrupted.

Companies and individuals can take proactive measures by limiting device usage on the water or at beaches to minimize disruptions to sea turtles.

From a conservation standpoint, it is now necessar to consider potential impacts of human activities on these different mechanisms. The areas that sea turtles keep returning to are important for those turtles, and these areas need strong protection and conservation.

18. CONCLUSIONS

What creatures! How important it is to conserve them

19. REFERENCES

1. Andersson, S.; J. Ornborg; M. Andersson (1998). "Ultraviolet sexual dimorphism and assortative mating in blue tits". Proceedings of the Royal Society B. 265 (1395): 445–50. doi:10.1098/rspb.1998.0315. PMC 1688915

2. Bright, Ashleigh; Waas, Joseph R. (August 2002). "Effects of bill pigmentation and UV reflectance during territory establishment in blackbirds" (PDF). Animal Behaviour. 64 (2): 207–213. doi:10.1006/anbe.2002.3042

3. Patricia Edmonds (2015). "The colourful Language of Chameleons". National Geographic Society Magazine March 2016 issue.

4. Anna Salleh (2023). abc.net.au/news/weird-animal-sounds-science-acoustics-fish-koala-lyrebird-frog-/102697402

5. "Tawny Frogmouths". Backyard Buddies. backyardbuddies.org.au

6. Sherman, Paul W.; Jarvis, Jennifer U. M.; Braude, Stanton H. (August 1992). "Naked Mole Rats". Scientific American. 267 (2): 72–79. Bibcode:1992SciAm.267b..72S. doi: 10.1038/scientificamerican0892-71 JSTOR

7. Sherman, Paul W.; Lacey, Eileen A.; Reeve, Hudson K.; Keller, Laurent (Spring 1995). "The eusociality continuum". Behavioral Ecology. 6 (1): 102-108. doi:10.1093/beheco/6.1.102

8. Welsh J (2011-10-12). "Naked Mole Rat Genome May Hold Key to Long Life". Human Health & Longevity. LiveScience, published October 13, 2011

9. Wang L, Cheung JT, Pu F, Li D, Zhang M, Fan Y (2011). "Why do woodpeckers resist head impact injury: a biomechanical investigation". PLOS ONE. 6 (10):e26490. Bibcode:2011PLoSO...626490W. doi:10.1371/journal.pone.0026490. PMC 3202538. PMID 22046293.

10. Puiu T. "Why woodpeckers don't get headaches". ZME Science. 23 March 2017

11. May PR, Fuster JM, Haber J, Hirschman A (June 1979). "Woodpecker drilling behavior. An endorsement of the rotational theory of impact brain injury". Archives of Neurology. 36 (6): 370–3. doi:10.1136/bjo.86.8.843. PMC 1771249. PMID 454236.

12. O'Shea, S.; K. Bolstad (2008). "Giant squid and colossal squid fact sheet". The Octopus News Magazine Online.

13. Froese, Rainer; Pauly, Daniel (eds.). "Family Synodontidae". FishBase. December 2008 version.

14. O'Corry-Crowe., G et al. Frontiers in Marine Science, published on February 27 2025.

Made in the USA
Coppell, TX
07 November 2026

11521264R00040